Hive

Suzanne Mercury

# Hive

Lily Poetry Review Books

Published by Lily Poetry Review Books
223 Winter Street
Whitman, MA 02382

https://lilypoetryreview.blog/

ISBN: 978-1-957755-32-8

Library of Congress Control Number: 2023941697

Cover design:  Martha McCollough

*For Sadie, Elizabeth, and Sarah*

*and for my teachers, the bees*

# Introduction

I wrote *Hive* as an act of magic, using a syllabic formula based on Agrippa's Magic Square for Venus. I composed it slowly over the course of winter and spring, finished it in May, then read it aloud while sitting among the hives while the bees swirled about me in the early summer heat.

Not a single sting: I think they approved.

Heinrich Cornelius Agrippa von Nettesheim (1486-1535), best known simply as Cornelius Agrippa, founded several secret magical societies and wrote a number of books on magic. He constructed seven different magic squares that he aligned with the seven planets, Saturn, Jupiter, Mars, Sun, Venus, Mercury, and Moon, that have become standards in ritual magic. Magic squares consist of an arrangement of numbers in the form of a square so that every row and column, plus both diagonals, will add up to the same number, in this case 49.

Each section in this poem has a syllabic structure that corresponds with the number in each square. For reference, I put the number of syllables in the margin next to each section. I found the syllable counting to be very liberating and physical. I counted the rhythms with my body, often walking and chanting the poem as I composed. The line breaks and arrangement on the page were determined by my natural breath.

The overall effect is a kind of episodic decentralized swerve that, I think, fits the personality and life cycle of the hive. For me, it fits the soul of the hive and what it feels like to live within it as a beekeeper.

Suzanne Mercury

# *Agrippa's Magic Square for Venus*

| 22 | 47 | 16 | 41 | 10 | 35 | 4  |
|----|----|----|----|----|----|----|
| 5  | 23 | 48 | 17 | 42 | 11 | 29 |
| 30 | 6  | 24 | 49 | 18 | 36 | 12 |
| 13 | 31 | 7  | 25 | 43 | 19 | 37 |
| 38 | 14 | 32 | 1  | 26 | 44 | 20 |
| 21 | 39 | 8  | 33 | 2  | 27 | 45 |
| 46 | 15 | 40 | 9  | 34 | 3  | 28 |

I look through                         parted

fingers to balance

                              the sun's
                                                                              ( 22 )

quintessence

in the bee's

                              embodied flight

Dirty alchemist—

       you are a planet       burned by sun :

              Inanna returned to you in a swarm

( 47 )      Trembling—

she shook the sun's head           coruscated with pollen

        and regenerated the world—

I remember flight :

[                    ]

[              ]

And the sunflower's                    ( 16 )

heliotropic

gaze—

The hive is utterance :

a chorus of wings

a shadow theatre of

backlit liminality

( 41 )                                    [                    ]

[                    ]

golden ratio of honeycomb :

[              ]                    dark red resin :

and date palms :

Talk to the bees

Go :

Tell them everything—

( 10 )

Anchoress :

Cave                    (and)                    queenlight

    rustle wings                              far-shining

     almond-eye                              thistle :

( 35 )

      you instruct me        *glisten this silver*

    (O rain!)

     *our wings        our wildflowers*

She is the

[                    ]

    [                    ]

                                                               ( 4 )

sun—

Earth bones    [                    ]

( 5 )                                    [                    ] veiled thorax—

The seed is sun

                                  in a bronze       wisp

that holds ten times its weight

in rainwater—                                           ( 23 )

[          ]

[          ]

                    Inanna takes flight—

The world reveals itself most

to travelers       lit up in the perspective

                     of fireflies

The hive is a meadow's blur :

weeds and wings

       and sun glare

( 48 )

[        ]

    [        ]

moths seeding themselves into

      this milkweed

        bugloss and vetch

Semaphore          with honeycomb

rainwater          sunpound

[          ] Shine :

                                        ( 17 )

        [                ] Speak :

                        Resonate :

In Napa Valley :

beekeepers said good-bye to their hives

and then fled the fires :

Returning, one found her hive

(42)       bearding beneath a tree root—

          still alive :

               a sheltered

                 *vibrato*

Inanna weeps

        pine resin :

          [       ]

                                    ( 11 )

A bee

captured—

The queen is a planet

              thinking herself into my palm

     *Alightment*

( 29 )

A liquid star     marked in blue

         too heavy to fly—

Hairy-legged bee :

I have so much to say to you—

        a root unfurled :

chlorophyll spreads through me                                  ( 30 )

      You are a doorway

           that stings me

Even so I say

( 6 )                    [          ]

                                    this

The hive is further away than I thought :

Bees coat               an operatic constellation                    ( 24 )

of pollen

Smoke spread                    and ash filled the valley—

     *The bees can't forage* :

A beekeeper said :

*knocked on each hive, ear pressed to them, feared the worst*

( 49 )

*Each hive responded—*

    *the bees were still there alive*

singing with an anguished

 roar—

I could not ever be less

than a flower now :                                    ( 18 )

a bee sits on my head

Maybe the earth is done with us :

                                    The anthropocene

( 36 )          is a pile of dead moths on the floor of the hive

                Time will shine through this cold wax

                the bones written—

A bee cycle—

a bicycle—        a bike                                ( 12 )

of bees

Legs coated in blue pollen :

( 13 )                    sunlight and

                                   cobalt squill

Black shine pollen     Scarlet pollen

  Gold

    and then dust-green    rough turquoise

             cell :      ( 31 )

      wind-caught  seed—

infinitesimal  enclosed

   in meaning—

( 7 )                    I've been stung twelve times today—

The hive is a blast-riot of

color-filled cells

( 25 )

      I love the black ones best :

           O scattered Spirit Mirrors—

Opium poppy        (Viper's Bugloss)        Horse Chestnut

Mountain Ash                        Cockspur thorn

Blackthorn        Moss        Meadowsweet

( 43 )                (Deadly Nightshade)

Hairy Willow                        Wild Cherry

Woodland Bluebell        Evening

Prim—rose

Hold me     (for me)          for one moment

             while Inanna

                                                       ( 19 )

                 unfurls herself

                         from her cell—

In the nimbus of           this gray morning

Apparitional Queen :

I am awake                for you

( 37 )

We are written in scent :

smoke      bergamot

cyperum           buds of myrrh

It was a defect

        a wound in the universe

that induced matter to gather        and produce

all this burr comb

                                                                                              ( 38 )

[        ]

[        ]

                    galaxy distribution

inside the hive—

( 14 )

If I could find anything blacker than

black      I'd use it

Honey is eternal :

found in graves          in clay jars

gold      as if new—                                              ( 32 )

You could eat it right there      with your fingers

standing there      at their door

(1)

[            ] [                ] [                  ]

[                ]    Heat

[                ] [                  ]

[                  ]

*Meditatio mortis*     dead layers

of leaves

       drones—

winter detritus and                    a continual crossing

The west     wind slips     silver     the margin

lacewings     lift     cordgrass     from blue

*Time*     *isn't*     *made*     *of anything*

slumgum     brood nest     honeycomb     queen flight

All of this     world:     shimmer planet     that I was

The sun is hive memory :

a levitating library                                    ( 20 )

of early nectar

The heart is sweet :

                    and it takes strength

( 21 )

to keep it that way :                              The steel strength

                    of hairy spiders

Shaking the bees from      their box      into the new hive :

   they swirl about                in the cloud of buzz

bumping against                                    bare hands

   and veil                                Lightness                    ( 39 )

    (sunlight and propolis)

     is all

My

cheek is covered

with

( 8 )

[                    ]

[          ]

pollen

The bees covered this dead mouse in propolis :

       methodical       they removed

    its hair first :

                                                    ( 33 )

[                ]

              O Queen Mouse!

You live                   eternal in black shroud sap—

( 2 ) *She* *lives*

The hive is Arkhē—

                from the Greek : an archive

a place from which everything

converges        cells

secrets

               and larvae

Enough for one day        these bees        tangled in my hair!

This one is lost :

                                        saddlebags filled with pollen

( 45 )

tarsals kicking against curl

buzzed        until I gently pulled her free :

                                *That way darling, that way—*

Plane of light        in the face-tar of trees :

        Propolis is beeswax combined with the

exudate from

                        pitch pine and spruce

                        fossil        heat        ( 46 )

                        and an accidental

                        lacewing—

        Who knows        how anything

                        happens?

The hive is a soft fissure

( 15 )                 in time :

                    reclaimed                    in nectar flow

*Insnar'd in flowers, I fall on grass :*

with a pollen heavy bee riding my hair

Is it ever not spring?                The sun sings in me                ( 40 )

Somewhere in the world                it must

be spring—

When you see this

( 9 )

your wing will kiss me

Why is this honey bright red?

The bees gathered its nectar from the maraschino cherry factory

in Redhook—

[          ]

[          ]

[          ]

[          ]

       Maybe you should not ask

( 34 )

( 3 )                    *Save*            *yourselves*

The hive is my soul school :

a wild book :

       fierce with self-propelled flowers

     in the marginalia                   ( 28 )

*Tell me—*

      *what—*

          *wrote this—*

## Endnotes:

"Time isn't made of anything" is a quote from Anne Carson's *Autobiography of Red*.

"Insnar'd in flowers, I fall on grass" is a quote from Andrew Marvell.

"If I could find anything blacker than black, I'd use it" is a quote from Joseph Mallord William Turner.

## Acknowledgements

My deepest thanks to Eileen Cleary– my publisher, dear friend, and kindred spirit– for her warmth, vision, and support in bringing this book into being, and to Martha Mccollough for her beautiful design and her feedback every step of the way.

I also owe a great debt of thanks to Kevin Gallagher who featured an excerpt from this poem in *spoKe* magazine, and to the Grolier Bookshop where I first read this poem to a non-bee audience. There are no words for how deeply honored I feel.

Many thanks to everyone who gave me feedback especially during the early days of writing this poem including Kevin McLellan, Darren Black, Josette Akresh-Gonzales, Cynthia Bargar, Annie Won, Maria Damon, Ruth Lepson, Christina Strong, Christine Jones, Andrew K. Peterson, and Ros Zimmerman,

Most of all, my heartfelt thanks to all of my fellow beekeepers, to the Boston Area Beekeepers Association where I first got started, and especially to Ang Roell of They Keep Bees for bringing together their magnificent summer Queen School where I learned so much about the practical magic of beekeeping.

And of course, always and forever, for the bees.

## About the Author

*Photo by Adrianne Mathiowetz*

Suzanne Mercury is the author of two chapbooks, *Sassafracas* (Xerolage 69), a collection of photographs of visual poems that she made out of scraps of dichroic glass (2018, Xexoxial Editions) and *Hand to Earth* (2019, Portable Press at Yo-Yo Labs). Her work has appeared in a variety of publications including *SpoKe, Truck, Summer Stock, Bombay Gin, Sonora Review, Arts & Letters,* and *Hayden's Ferry Review,* as well as in the anthologies *Let the Bucket Down* and *The Wisdoms of the Universes in a Single String of Letters.* A graduate of Smith College and Syracuse University's MFA program in creative writing, she lives in the greater Boston area where she creates sustainable gardens and keeps bees.

www.ingramcontent.com/pod-product-compliance
Lightning Source LLC
Chambersburg PA
CBHW081005140626
46546CB00019B/3435